What is a Trace Fossil?

D1594970

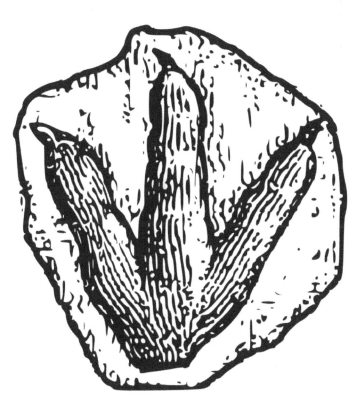

A Coloring Book by
The Georgia Mineral Society, Inc.

Written by Lori Carter

Copyright © 2014
The Georgia Mineral Society, Inc.

A completely volunteer, non-profit 501(c)(3) organization dedicated to
educating the youth of the state and members of our society
in the field of Earth Sciences

All proceeds above cost will be used to further education in Earth Sciences
and to support the educational efforts of the Georgia Mineral Society.

This edition published by:

The Georgia Mineral Society, Inc.
4138 Steve Reynolds Boulevard
Norcross, GA 30093-3059
www.gamineral.org

ISBN: 978-1-937617-09-7

What is a trace fossil?

Is a trace fossil a fossil that was traced?

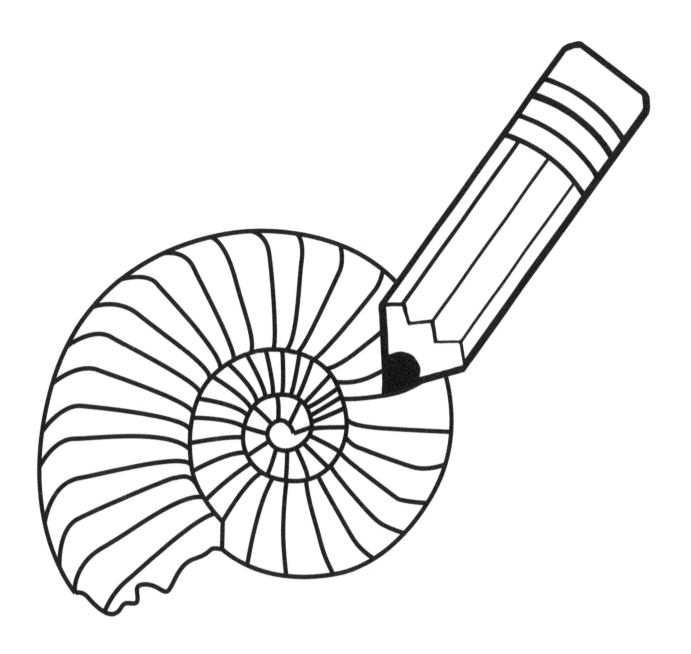

A trace fossil is a fossil that can show the activity of something that lived a long time ago. For example, walking, feeding, digging, pooping, resting, or growth.

The Georgia Mineral Society, Inc.

When the trace was made it might have been squishy...

The Georgia Mineral Society, Inc.

The Georgia Mineral Society, Inc.

...but with the right conditions and a lot of time...

Mon	Tue	Wed	Thr	Fri	Sat	Sun
			1	2	3	4
5	6	7	8	9	10	11
12	13	14	15	16	17	18
19	20	21	22	23	24	25
26	27	28	29	30		

...the trace can be preserved in stone

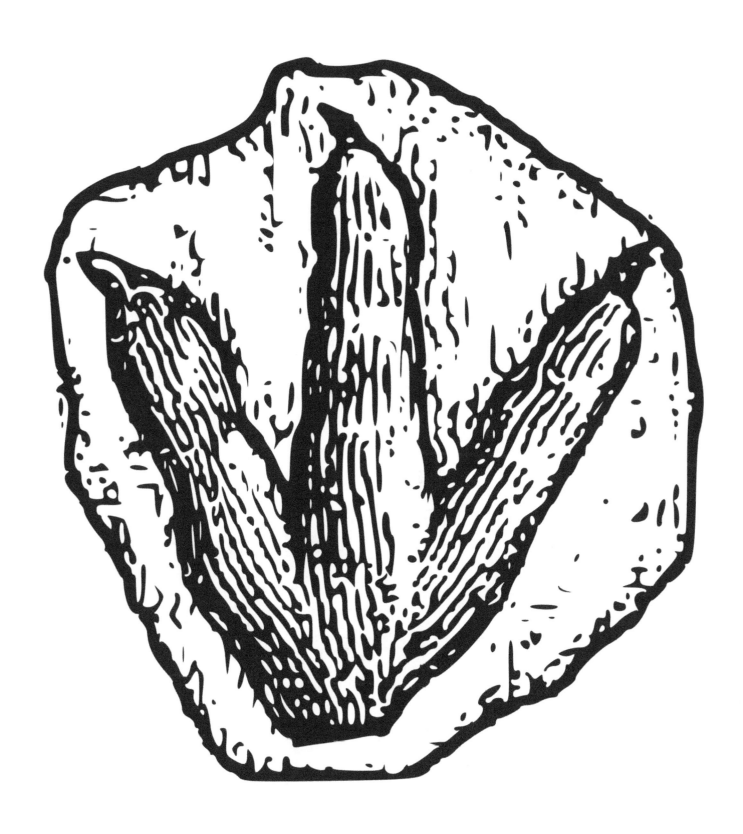

What kind of trace fossils have been found?

A footprint can be a trace fossil

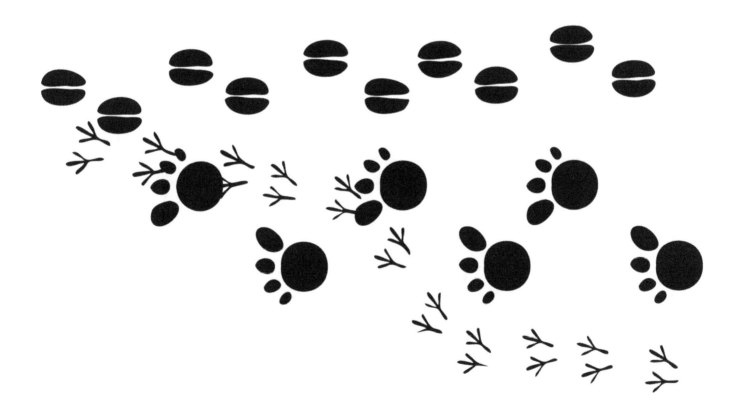

When a footprint becomes a trace fossil it is called a "track"

Lizards left tracks that became trace fossils

Dinosaurs left tracks that became trace fossils

Birds left tracks that became trace fossils

Insects left tracks that became trace fossils

Even creatures that don't exist anymore, like trilobites, left tracks that became trace fossils

Fossilized poop is a trace fossil too

The cast of the path of root growth is another kind of trace fossil

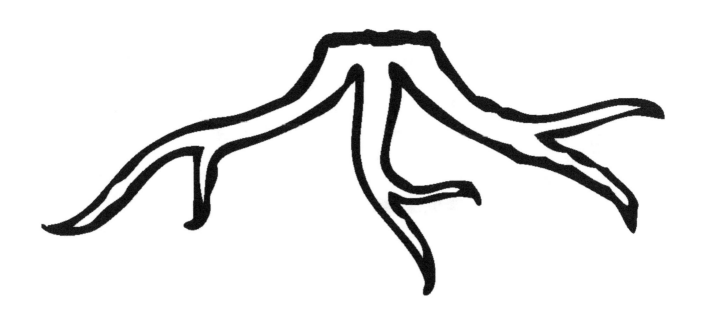

Does that mean petrified wood is a trace fossil?

No, petrified wood is preservation of an <u>organism</u> (the wood).

A trace fossil is preservation of an organism's <u>activity</u> (the root growth).

So, a bite mark on a fossilized bone would be a trace fossil...

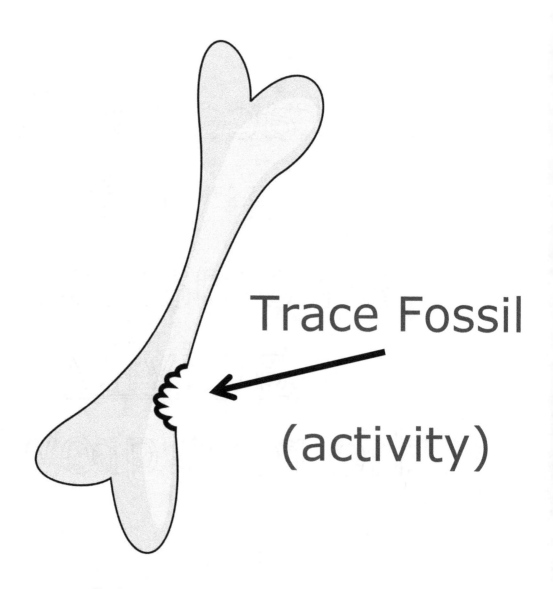

Trace Fossil

(activity)

...but the fossilized bone itself would not be a trace fossil

Fossil

(organism)

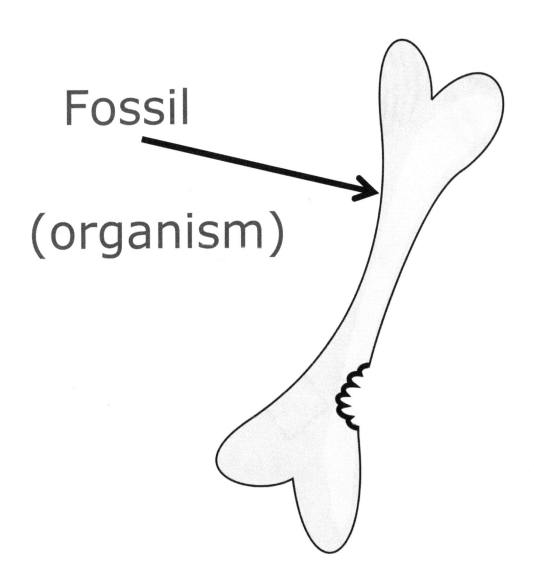

Fossil
(animal)

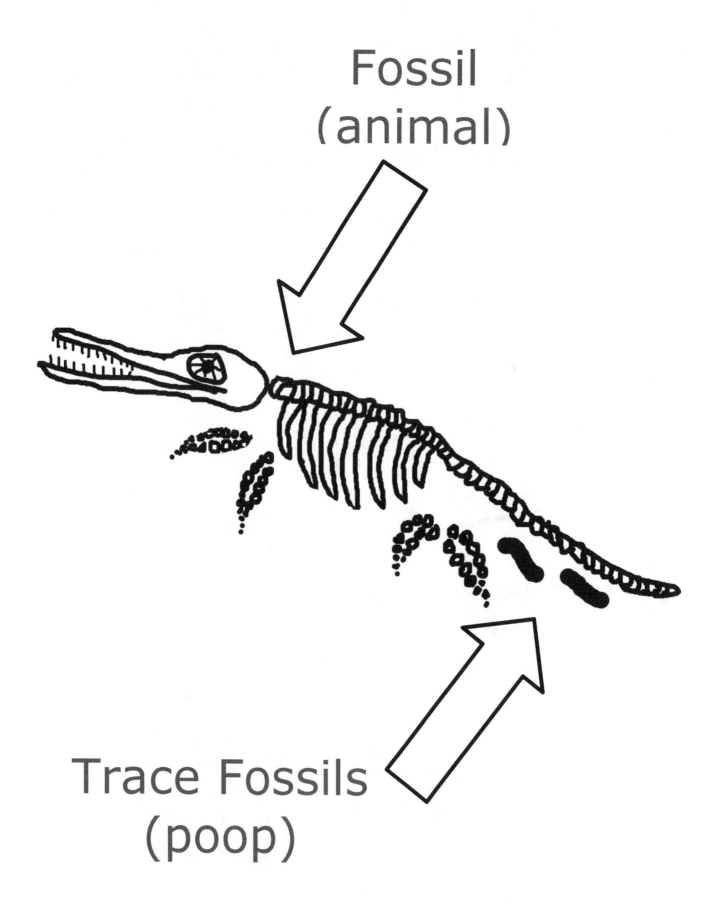

Trace Fossils
(poop)

Fossils
(bones)

Trace Fossils
(tracks)

Another trace fossil is a burrow made by something digging

The burrow may be preserved as a trace fossil, but the creature that dug the burrow may not be preserved with it

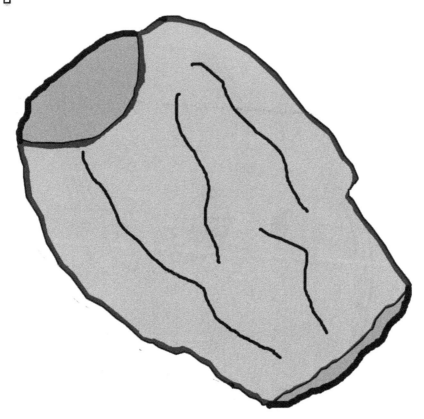

Another trace fossil is a spot where something stopped for a moment. It is called a "resting" trace.

Let's play a trace fossil game!

Find the trace fossils!

Did you find all of the trace fossils?

Fun Facts About Trace Fossils!

Mud cracks, mud ripples, and rain drop impressions are not trace fossils. They are not even fossils. They are sedimentary structures.

The study of trace fossils is called "ichnology" (ick-nahl-uh-jee) and people who study trace fossils are called ichnologists.

Trace fossils have their own names that are not necessarily related to the organism that made the trace because the source of the trace may not be known.

The difference between a trace fossil and a body fossil is that a trace fossil is from an organism's <u>activity</u> and a body fossil is from an organism's <u>body</u>, even if it is a mold or cast.

So, a root cast can be a trace fossil or a body fossil. If the cast is from a root that grew and left a void, that would be a trace fossil because it occurred from biological activity, in this case the growth of the root.

If the cast is of the root itself, it would be a body fossil, because it is a geologic record of the body of the organism, in this case the actual root.

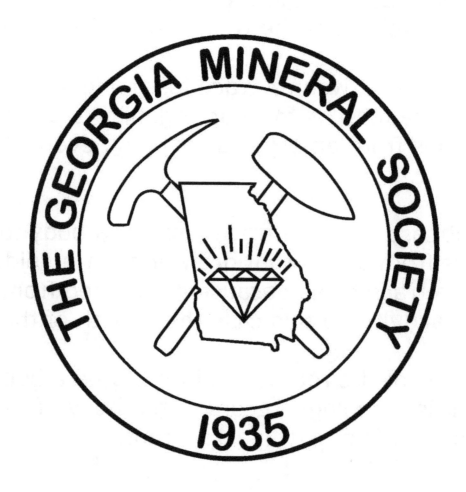

The Georgia Mineral Society, Inc.